Auta Paulina da Silva Oliveira
Márcia Adelino da Silva Dias
Alef Nunes da Silva

Interdisciplinary approach to teaching biology: a teaching experience

Auta Paulina da Silva Oliveira
Márcia Adelino da Silva Dias
Alef Nunes da Silva

Interdisciplinary approach to teaching biology: a teaching experience

Pedagogical interventions

ScienciaScripts

Imprint

Cover image: www.ingimage.com

This book is a translation from the original published under ISBN 978-613-9-68070-2.

Publisher:
Sciencia Scripts
is a trademark of
Dodo Books Indian Ocean Ltd. and OmniScriptum S.R.L publishing group

120 High Road, East Finchley, London, N2 9ED, United Kingdom
Str. Armeneasca 28/1, office 1, Chisinau MD-2012, Republic of Moldova, Europe
Printed at: see last page
ISBN: 978-620-8-19334-8

INDICE

CHAPTER 1

PERCEPTION OF THE ENVIRONMENT AND ANALYSIS OF ENVIRONMENTAL EDUCATION IN AN INTERDISCIPLINARY WAY FOR HIGH SCHOOL STUDENTS

Auta Paulina da Silva Oliveira[1] Alef Nunes da Silva[2]

Màrcia Adelino da Silva Dias[3]

Summary: The work was carried out at the Sólon de Lucena State School of Elementary and Secondary Education, located in the city of Campina Grande, PB. A 2nd year high school class was chosen as participants. The need to develop Environmental Education (EE) is affecting all educational institutions, including schools, which are training citizens. It is necessary to work on the environment so that students can be mature enough to deal with environmental problems. A key factor is the way in which it is taught, so teaching methodology and teacher training are of great importance. The aim of this work is to analyze students' environmental perceptions in order to raise their awareness and examine whether they receive environmental education in an interdisciplinary way. The methodology used with the students was the application of semi-structured questionnaires, construction of a mental map of environmental perception, explanation of content with the help of technological resources, presentation of posters and a dynamic, with the teachers only the application

[1]Master's student in the post-graduate program in agricultural sciences at the State University of São Paulo. Paraiba;[2] Graduated in Biological Sciences at the State University of Paraiba;[3] PhD Professor at the Department of Biology, State University of Paraiba.

of a semi-structured questionnaire. The lack of interdisciplinary environmental education was noted, but the conclusion of the first stage of the project resulted in a contribution to changing students' perceptions, achieving the objectives of the work.

KEYWORDS: Environmental education; Environmental perception; Environment; Interdisciplinarity.

INTRODUCTION

In view of the environmental problems that have arisen in recent centuries, such as natural disasters, droughts, compromised biodiversity and the others that have plagued the global population, it is essential to work on a theme in schools that prepares students to face them, so that they are the protagonists of an attempt to change them through simple everyday actions. "In the 20th century, the environmental effects caused by various human actions became increasingly significant." (REBOLLAR, 2009, p.174).

The environmental dimension is not only related to theoretical and scientific concepts, it is essential to raise awareness. It is necessary to innovate didactics in order to work on this theme, always seeking to move people, promoting new strategies so that transformation can actually take place (ROSA, 2009), so environmental education (EE) only shows results after the cooperation of everyone involved in the work

In order to work on environmental education, we must first get to know the environmental reality of the students, the community, the teachers and everyone involved, since a change in perception only occurs when it is possible to see an environmental problem nearby. Teachers still lack information and their didactics may not meet the expectations of how to work with environmental education.

Environmental education is undoubtedly the tool for changing human thinking and attitudes (BEZERRA and GONÇALVES, 2007).

According to CARVALHO (2001), environmental education has been incorporated as an innovative practice in different areas. The rapid environmental changes resulting from human activities threaten life, the propagation of species and, consequently, may be related to the way people understand the environment (BEZERRA and GONÇALVES, 2007), so analyzing the environmental perception of the group to be studied is fundamental for structuring methodologies and the way in which the problem should be approached.

Environmental education has a heterogeneous universe of intentions. Teachers must seek out socio-educational forms, change their pedagogical methodologies and, above all, reformulate their ideals in order to work in this area, always aiming to achieve the desired environmental change (CARVALHO, 2001), and also states that environmental education has been given this name because it is understood as the differential, which establishes an understanding of the relationships and interactions in the natural and social base.

Through it, we can understand that the natural ecosystem is actually a space of socio-environmental relationships, which over the years has been shaped and modified by the tensions and conflicts arising from society. For a number of reasons, understanding and working with the environment is not an easy task; it requires knowledge and didactics that go beyond the traditional paradigm of education.

Due to the lack of continuing training for teachers on the subject of the "Environment", given that these issues should not only be addressed at events or on specific days, there is a deficit in education. It is therefore a challenge to propose that institutions invest in training multiplier agents in environmental education and in changing the curriculum of academia to train teachers, "who, however, are not receiving sufficient training to break the chains that bind them to a conservative educational perspective" (ROSA, 2009, p. 456). According to the author:

The realization of environmental education requires the use of new methods and practices in the educational system, it requires a curriculum that enables the acquisition of the school's identity and the valuing of educators, as well as a methodological approach that favors the establishment of a critical and reflective dialogue... (ROSA, 2009, p.458)

The main objective of this study was to analyze the environmental perception of 2nd year high school students at the Solon de Lucena State School of Elementary and Secondary Education, located in the city of Campina Grande, Paraiba; to raise their awareness and examine whether they are receiving environmental education in an interdisciplinary way.

METHODOLOGY

This work was carried out at the Solon de Lucena State School of Elementary and Secondary Education in the municipality of Campina Grande-PB, and is the conclusion of the first stage of the biology subproject, developed through the PIBID (Institutional Program of Teaching Initiation Scholarships) of the State University of Paraiba (UEPB).

The school is located in the Centro district, close to the central business district of Campina Grande. It has elementary and high school grades (6th grade to 3rd grade), working all shifts (morning, afternoon and evening).

The methodology is similar to that used by BEZERRA (2008), (application of semi-structured questionnaires and construction of a mental map of environmental perception) and ROSA 2009 (semi-structured questionnaires and meetings).

The methodological processes are based on the construction of a mind map, the development of posters, discussions on the topic, dynamics, as well as the use of technological resources for slide presentations (projector, speaker). BEZERRA (2007) states that using a mind map is a way of monitoring something, and can be used as a method for analyzing each individual's conceptions and visions.

The meetings with the class took place every other day during the 2015.1 semester, more precisely from March to June 2015. During this period there was a strike by primary school teachers in the state of Paraiba (it lasted throughout April), which reduced the number of meetings. Initially, the students were asked to build a mind map so that they could draw pictures of the environment. In subsequent meetings, the students developed posters describing the environmental problems surrounding their city and neighborhood.

The environmental theme was addressed through slide shows, as well as discussions about the issues, video presentations, the students' opinions on the problem and a dynamic game about solid waste. The students also answered a semi-structured questionnaire asking about their gender, age, neighborhood, the contribution of the research to their lives and which subjects dealt with environmental education. Teachers were also asked to complete a semi-structured questionnaire with 10 questions:

1st Gender MAS ( ) FEM ( );

2nd Age__;

3° Time in the job__;

4° Education; ( )Bachelor's degree, ( ) Bachelor's degree with specialization, ( ) Master's degree, ( ) Doctorate, ( ) Post-doctorate;

5° What is your area of expertise, and which subjects are you responsible for mentioning?

6° Do you believe that environmental education should be taught in all areas of knowledge?

7° If you work from the perspective of environmental education and its interdisciplinarity, what didactics do you use so that the content is accepted and absorbed by the students?

8° Do you believe that your teaching helps you to interconnect subjects, especially those related to Environmental Education?

9° In view of the environmental problems plaguing our city, what activities do you promote in the school to help students become environmentally aware?

10° In your opinion, should Environmental Education be incorporated as a subject?

RESULTS AND DISCUSSION

EE should be understood as part of the study of environmental issues and awareness-raising is the key element, which can be developed through schools, governmental or non-governmental organizations, informally or formally (BEZERRA E GONÇALVES, 2007).

EE is one of the fundamental tools for encouraging research as a way of instigating the students' curiosity, without dispensing with the teacher's position, who is responsible for driving the students' action, always aiming at the reconstruction and construction of knowledge (ROSA, 2009), which takes place through exchanges.

The project involved the second year class (A), made up of 20 students from the public school. The results of the questionnaires showed that the students live in different neighborhoods in the city of Campina Grande (Alto Branco, Catolé, Santo Antônio, Bairro das Naçôes, Palmeira, José Pinheiro, Conceiçào and Centro), with the exception of one student who lives in the city of Massaranduba. The diversity of neighborhoods contributes to a differentiated view of the problems in each locality. The students' ages range from 15 to 19, and the class is divided into 12 males and 8 females.

A mind map was proposed, since it is essential to identify the perception of the individuals involved (ROSA, 2007), so the students presented the environment in the form of drawings. The mind map is a methodology that is suitable for analyzing environmental perception due to its symbolic representation, portraying the vision of an environment as it should be or as it really is for each person (BEZERRA, 2008).

"Perceptions of the environment are relative and always fragmented because they are filtered by social systems, customs and religions." (REBOLLAR, 2009, p.174). In this way, they present a naturalistic view (figure 1a) and a globalizing view (figure 1b), (BEZERRA E GONÇALVES, 2007).

Some students saw the environment as consisting only of nature, animals, trees and bodies of water, while others took a more global view, showing that the environment is made up of all beings, including human beings, emphasizing through their drawings the impact of man's actions on the environment.

1a)

1b)

Figure a and b. Naturalistic view (a) and globalizing view (b). Source: Author, 2015.

The drawings were analyzed with the class and in order to emphasize the importance of building the map, there was an exchange of knowledge on the subject, emphasizing that everything is the environment. The results of the mind map are shown in Table 1.

Table. Analysis of the mind map according to Rosa et al. (2009).

ENVIRONMENTAL CONCEPTUALIZATION THROUGH STUDENTS' MIND MAPS	
CATEGORY	PERCENTAGE
Naturalist Vision	46.67 %
Global vision	53.33%
Total:	100%

According to the data obtained through the students' perceptual analysis of the concept of the environment through their drawings, the lack of environmental education was ascertained. The students were shown images of the problems through videos and slide shows, in which the Bodocongó reservoir (located in one of the city's districts, close to the State University of Paraiba - UEPB) was cited as one of the city's main impacted ecosystems.

With this in mind, the students also developed posters with the environmental problems they believed to be aggravating not only in the city but also globally. These posters were shared with the class (figures 2a and 2b) and dealt with different topics, such as water scarcity, noise pollution and deforestation.

Noise pollution is one of the problems that is not often dealt with in schools, but it was well presented by a group of students, citing examples of noise from the city's factories and from the central business district, which is located close to the school.

Figure. 2a and 2b. Socialization of the posters on the proposed theme. Source: Author (2015).

In the light of the results of the questionnaire, regarding the importance of the project, the majority of the students claimed that the project was very good and that there had been a change in their way of thinking and acting, but a small percentage said that there had been no change or that the change had been partial, since they already had knowledge of the subjects covered (table 2).

Table 2 - Results of the questionnaire regarding the change in students' perceptions with the development of the project, showing the category and percentage, corroborating Rosa et al., (2009). Source: Oliveira, 2015.

CATEGORY	PERCENTAGE
Occurrence of change	83.34%
Partial Change	8.33%
No change	8.33%
Total	100%

The students were also put to the test with a solid waste dynamic. They took the waste out of the garbage can in front of the classroom, and when they saw the room full of scattered waste, they started making comments asking why the room was dirty, showing that they were annoyed.

When they settled down, we explained why the room was the way it was and the relationship between the dynamics of solid waste and the topic we were working on. As well as differentiating between the concepts of garbage and solid waste, there was a discussion about selective waste collection at the school, and it was noted that selective waste collection does not take place correctly at the school in question. The waste containers are scattered around randomly with their lids changed and the students are not instructed to dispose of their waste in the appropriate place.

Concluding on the data obtained from the students' questionnaires, it can be seen that only three disciplines work from an interdisciplinary perspective (biology, geography and history). This is similar to the analysis by ROSA (2009), who also presented only three disciplines that worked with EE: chemistry, biology and geography.

The data differs from that presented in the questionnaires answered by the teachers, who presented physics as a subject that approaches environmental education in an interdisciplinary way. According to REBOLLAR, 2009, in order for environmental education to take place, it is extremely important for several researchers from different areas to interact, which is why it is vigorously interdisciplinary.

Ten teachers were interviewed using a semi-structured questionnaire. They taught in the school's different shifts, in regular classes and in the EJA (Youth and Adult Education) project. Among the teachers who didn't answer, some refused due to lack of time and others handed in blank questionnaires. Based on the interviewees, it was observed that they work in the areas of Portuguese language, mathematics, history, physics and biology. The results of the questionnaires submitted to the teachers are summarized in Table 1.

Of the teachers interviewed, only 40% had an interdisciplinary perspective on working with environmental education, working in the areas of physics, biology and history. These teachers believe that their didactics help them to interconnect content and the theme of "the environment". It's worth noting that the teachers who didn't answer the questionnaire were from other disciplines such as chemistry and geography.

Chart 1- Results of the questionnaire applied to teachers at the Solon de Lucena State Elementary and Secondary School in 2015. Source: Oliveira, 2015.

Questions	**Results**
Gender	5 male and 5 female teachers.
Age	Range from 25 to 55 years old.

Level of Training	Bachelor's degree; Bachelor's degree with specialization and Master's degree.
Length of service	Range from 3 to 30 years of experience.
Areas of activity	Portuguese, mathematics, history, physics, biology.
An interdisciplinary approach to working with environmental education	Only 40% of teachers claimed to work with EE in their school practices.
Implementing environmental education as a subject	40% of teachers support the implementation; 40% disagree and 20% did not answer.

The information obtained from students and teachers raises questions about the occurrence of environmental education in its interdisciplinary form, since teachers approach physics as an interdisciplinary practice for working with environmental education, while students do not recognize it.

FINAL CONSIDERATIONS

After the end of the first stage of the project, it was concluded that the school lacks the ability to work with environmental education in an interdisciplinary way, but the project has had positive results. There was an improvement in the students' concept of the current environmental problems we are experiencing. An alternative for environmental education to become more widespread in schools would be better academic training for education professionals, who, together with the family, will be responsible for forming new, aware citizens.

CHAPTER 2

EXPERIMENTAL ACTIVITY AS A TEACHING AID IN THE TEACHING OF BIOLOGY: AN EXPERIENCE REPORT

Auta Paulina da Silva Oliveira[1]

Alef Nunes da Silva[2]

Màrcia Adelino da Silva Dias[3]

Abstract: It is extremely important to develop experimental activities for primary school students in various subjects. Experimental activities promote greater interaction between students and teachers, which can lead to new planning and the development of new strategies. With this in mind, the aim of this paper is to discuss the experience gained in a public school in the city of Campina Grande, through the development of a pedagogical intervention that enabled an experimental activity in the classroom with high school students. Four simple experiments were developed as a complement to some subjects already covered in the classroom relating to the environment. Through the experimental activities, it can be seen that students don't participate in experimental practices very often and that this type of activity is very attractive, as it allows students to get involved and ask questions, contributing to an interactive practice.

KEYWORDS: Experimentation; Science; Reuse.

INTRODUCTION

Faced with educational paradigms and the mutual discouragement of teachers and students, it is interesting to develop projects, practical and recreational activities and, above all, to bring students closer to scientific knowledge. In view of the educational context in which we live, we cannot remain submissive to a traditional and banking education, with the teacher being the central figure of education and the students passive receptors (FREIRE, 1987; MIZUKAMI, 1986).

The development of experimental activities for primary school students in various subjects is extremely important. All areas of knowledge are characterized by being transdisciplinary, adding knowledge and information from all areas, which, if approached in the right way, will provide students with an easier dialogue, facilitating their teaching and learning, as well as helping to break down the educational paradigms listed by Mizukami (1986). In this way, teaching biology requires the teacher to have ownership, maturity and improved teaching methodologies.

Experimental activities promote greater interaction between students and teachers, which can lead to new planning and the development of new strategies that result in greater assimilation of science content (MORAES et al. 2000). However, the lack of continuing training for teachers, the lack of equipment in some schools, the lack of time for teachers to dedicate, and the lack of training offered by the school in handling equipment and the wrong choice of teaching material, make education professionals unmotivated and often only passively transmit the content (PAGEL, 2015).

Experimentation plays an essential role in science teaching and can have positive results or not, depending on the degree of interest of the students and the way it is approached. Bringing students closer to the common experimental environment, the laboratory, is essential in order to provide a link between scientific concepts, mental models and students' prior knowledge. This type of activity seeks not only to demonstrate phenomena in a ready and definitive way, but also to arouse curiosity and interest among students in seeking out new information (NEVES et al, 2006).

The general aim of any practical lesson, whether experimental or not, in biology teaching is to result in science learning. However, in order to learn science, it is essential to practice science and this practice will result in reflection (MORAES et al. 2000). Scientific knowledge can be approached in different ways, through experimental practices, playful activities such as games, gymkhanas, dynamics, among others.

In view of the reality of many Brazilian schools which do not have laboratories, which have unprepared professionals or where the space for experimentation is used for other purposes, there is a need to create new methodologies and use creativity by the teacher in partnership with the students.

Experiments can be carried out in the laboratory, in the classroom or outside it, even using low-cost materials, which allows for reciprocal creativity between students and teachers (MORAES et al, 2000). From this perspective, this paper aims to address the experience of a public school in the city of Campina Grande, through the development of a pedagogical intervention that enabled an experimental activity in the classroom with high school students, which promoted a greater association of the knowledge covered in the subject of biology during the first two months of the 2016 school year.

METHODOLOGY

This experience was carried out at the Solon de Lucena State School of Elementary and Secondary Education in the municipality of Campina Grande-PB, and is the conclusion of the first stage of the biology subproject, developed through the PIBID (Institutional Program of Teaching Initiation Scholarships) of the State University of Paraiba (UEPB). The school is located in the Centro district, close to the city's central business district, and has elementary and high school grades (6th grade to 3rd grade), operating in all shifts (morning, afternoon and evening).

The class chosen for the PIBID educational intervention is a first-year high school class, whose students range in age from 15 to 17 and live in the city of Campina Grande-PB. Prior to the activity, classes given by the teacher responsible for biology were observed for two months.

An analysis was carried out through observation, which revealed that the teacher did not develop experimental lessons. A pedagogical intervention was then developed using low-cost materials for experimental practice.

Four simple experiments were developed to complement some of the subjects already covered. The environment chosen to conduct the experiments was the classroom itself, since there is no definitive space for practicing science, and it can be developed inside or outside a simple classroom (MORAES, et al. 2000). It's worth noting that the choice of environment was not random and purposeful, since the school's science laboratory functions as a classroom for special students.

The following materials were used for the activity: a glass dish (one large and one small), a glass beaker, a candle, disposable cups, a match, a spoon, sand, water and a pet bottle. The class was divided into four groups, which were responsible for observing the experiments, and then participating in a discussion about the activity, addressing which subjects that had been studied previously could be related to the experiment.

RESULTS AND DISCUSSION

The experimental activity that made it possible to develop this experience report covered various subjects relating to the content covered in the textbook and used by the teacher responsible for the subject "Biology" in the first two months of the 2016 school year. The experimental activity was simple and low-cost materials were used to carry it out. At the start of the intervention, the students asked: *"Teacher, are you going to set fire to this?"; "Is there going to be an explosion?".*

These statements demonstrate the students' initial view of experimentation, i.e. their mental models. In this sense, Errobidart et al. (2013) state that mental models relate the sequence of the discourse, event or situation, including the objectives involved and the implicit knowledge that the observer uses to understand them (p.441).

The experiments chosen were adapted from those available on the "Experimentoteca-CDCC" (a website whose experiments are aimed at primary and secondary schools, provided by USP - University of Sao Paulo) and from the mental models of the undergraduate student who developed the project and as a conclusion to the first stage, the pedagogical intervention.

The aim of the first experiment was to demonstrate the importance of oxygen (O_2) in the combustion process and the release of carbon dioxide in its absence, using a plate, a candle, a beaker and a match. The procedures are shown in figure 1, in three steps. The plate was used as a support for the candle, the phosphorus to produce the flame and the beaker as an indispensable tool in the process in which the flame is extinguished and carbon dioxide (CO_2) is released.

Figure 1. First experiment, demonstrating the importance of oxygen for the combustion process and the release of carbon dioxide (CO_2). The image shows the three steps, respectively. Source: Oliveira, 2016.

The second experiment aimed to demonstrate the process of soil erosion. A pet bottle cut in half with sand, a plate and a disposable cup with water were used. The procedures were as follows: the bottle with the sand was tilted under the plate and then the water was gradually poured over the sand, allowing it to drain away.

Through this process, the students were able to draw a parallel with what happens in nature, through floods that cause erosion by carrying away all the sediment when the soil is unprotected or due to anthropogenic actions. The stages of the experiment are illustrated in figure 2.

Figure 2 Second experiment. Demonstration of soil erosion. Source. Oliveira, 2016.

To carry out the third experiment, the materials used were three disposable cups, sand and water. The aim of this experiment was to demonstrate how the process of water pollution takes place and that this process is always cyclical, as it directly or indirectly reaches the initial point of pollution. The three glasses contained water, and initially a little sand was placed in one of the glasses, turning the water dark. The dark water was then transferred to the other glasses. The stages of this experiment are illustrated in Figure 3.

Figure 3. Third experiment, demonstrating water pollution along water bodies. Source: Oliveira, 2016.

The fourth and final experiment aimed to demonstrate the process of water evaporation in the soil and that unprotected soil is vulnerable to disasters and can become infertile because it has nowhere to get its nutrients from. A spoon, a large plate, a small plate, sand, a candle and a match were used.

The sand was placed in the spoon, then the candle was lit and the spoon was positioned just above the flame. The larger plate was left on the table to support the candle and the smaller plate was placed above the spoon to visualize the water droplets formed by the evaporation of the soil. The processes described are shown in figure 4.

Figure 4 - Fourth experiment illustrating soil water evaporation. Source: Oliveira, 2016.

The experiments listed above prove that it is possible to develop experimental activities even in the absence of a science laboratory. However, it is more convenient and practical for science and biology teachers to use theoretical lectures. Teaching and learning biology requires dynamism to make the teaching-learning process a reality, which is why practical and experimental lessons have great pedagogical potential for transmitting and obtaining scientific knowledge (PAGEL, 2015).

FINAL CONSIDERATIONS

It is essential to use differentiated activities for primary school students, since it is at this stage of their studies that they will be deciding which professional career they will choose, and teachers may or may not be a source of inspiration for this decision.

Teaching methodology and practices can be directly related to the demotivation that is so prevalent among primary school students in all schools, especially public schools, due to their lack of funding.

The teaching-learning process in science and biology should be enjoyable and attractive, so experimental activities work as a differentiated and attractive teaching tool. This is a means of motivating students who are nearing the end of high school.

Through the development of an experimental activity with students in the first year of secondary school at a public school in Campina Grande, it can be seen that they do not participate in experimental practices very often and that this type of activity is very attractive, as it allows students to get involved and ask questions, contributing to an interactive practice.

It is worth mentioning that this experience report was developed from a pedagogical intervention developed through PIBID, a program of great relevance for initial teacher training, as it promotes this approach to the classroom, making it possible to assist in the training of a teacher-researcher.

CHAPTER 3

APPROACH TO ENVIRONMENTAL EDUCATION FOR HIGH SCHOOL STUDENTS

Auta Paulina da Silva Oliveira[1] Alef Nunes da Silva[2] Milena Maria de Luna Francisco[3] Màrcia Adelino da Silva Dias[4]

Abstract: Environmental education is an informative and formative process for individuals, developing skills and changing attitudes towards the environment, making the educational community aware of its global reality. The work aims to analyze the environmental perception of students; work on environmental education with the development of activities and in order to provide the students involved with awareness through activities with a sustainable vision, methodology was used the application of a questionnaire, a dynamic on solid waste and the construction of posters. The students demonstrated some environmental problems, highlighting the most noticeable in the city of Campina Grande, PB. It can be seen that the students have two views of the environment (naturalist and globalist) and that the school lacks interdisciplinary environmental education.

KEYWORDS: Environment; Environmental problems; Awareness.

INTRODUCTION

Environmental problems are becoming more worrying every day and "in the 20th century, the environmental effects caused by various human actions became increasingly significant." (Rebollar, 2009, p.174). Thus, the environmental dimension is not only related to theoretical and scientific concepts, it is essential to raise awareness through practical methods.

"Environmental education is an informative and formative process for individuals, developing skills and modifying attitudes towards the environment, making the educational community aware of its global reality." (Soares, 2007 p.5). Working on environmental education is no easy task and is undoubtedly the tool for changing people's perceptions (Bezerra and Gonçalves, 2007).

It is necessary to innovate didactics to work on this topic, always seeking to move people, promoting new strategies, so that transformation actually occurs (Rosa, 2009), so environmental education (EE) only shows results after the cooperation of everyone involved in the work.

EE presents a heterogeneous universe of intentions, teachers must seek socio-educational forms, change their pedagogical methodologies and especially reformulate their ideals to work on this aspect, always aiming to achieve the desired environmental change (Carvalho, 2001), he also states that environmental education received this denomination because it is understood as the differential, which establishes an understanding of the relationships and interactions in the natural and social base.

Due to the lack of continuing training for teachers on the subject of the "Environment", which should not be addressed only at events or on specific days, there is a deficit in education, and so there are barriers to proposing activities aimed at raising environmental awareness.

With a view to environmental problems, the aim of this work is to analyze students' perceptions of the environment; to work on environmental education through the development of activities and with the aim of raising awareness among the students involved through activities with a sustainable vision, based on the perspective of reducing, recycling and reusing; to examine whether students receive environmental education in an interdisciplinary way and thus develop methodologies so that the activities developed obtain positive results.

METHODOLOGY

This paper reports on the activities developed through the PIBID (institutional program of teaching initiation grants) of the Biological Sciences course at UEPB (State University of Paraiba). The project involved the second year class (A) of the Solon de Lucena State School of Elementary and Secondary Education, located in the Centro district of the city of Campina Grande-PB. The methodology was similar to that used by Bezerra (2008) and Rosa (2009), with questionnaires and meetings.

A solid waste dynamic and the construction of posters were also used as methodology. The meetings were held every other day from May to June 2015. During the meetings, the students were subjected to a dynamic on solid waste and made posters, which were shared with the class.

To conclude this stage of the project, the students were given a questionnaire with simple questions asking them to provide information on their gender, age, neighborhood and which subjects deal with environmental education in an interdisciplinary way. The diversity of the neighborhoods may be a factor that contributes to the students' different perceptions.

RESULTS AND DISCUSSION

The class participating in the project at the aforementioned public school is made up of 20 students, divided into 12 males and 8 females, who live in different neighborhoods in the city of Campina Grande-PB (Alto Branco, Catolé, Santo Antônio, Bairro das Naçôes, Palmeira, José Pinheiro, Conceiçao and Centro).

According to Rosa's (2007) definition, the students approached a naturalistic view and a globalizing view through a dynamic on solid waste (Figurei). To develop the dynamic, we used the waste from the class garbage can. Before the students arrived, we scattered all the waste around the classroom, which made the students feel uncomfortable.

This led to a discussion about selective waste collection at the school and the use of solid waste for other purposes, such as making decorative items and utensils, for example.

Through the discussion generated by the dynamics, the students made posters (figure 2) and presented them to the class as a way of raising awareness, which presented different themes such as: noise pollution, deforestation and water scarcity.

Through their posters, the students demonstrated the environmental problems closest to them, highlighting the most noticeable in the city of Campina Grande, PB. A rotation of presentations was made using the posters made to illustrate the drawings, and through orality they addressed issues demonstrating safety and concern about the situation of the city in various places.

1)

2)

Figure 1 and 2: Illustration of the solid waste dynamic (1). Illustration of the socialization of the posters (2). Source: Oliveira, 2016.

Among the problems presented, one group stood out for addressing the issue of noise pollution and water scarcity. The groups emphasized the rationing of water in the city and the noise from the factories as one of the main means of causing hearing problems among factory workers.

According to the students' answers in the questionnaire, it can be seen that only three subjects work from an interdisciplinary perspective: biology, history and geography, which shows the lack of interdisciplinary environmental education. It was found that the students vary in age from 15 to 19 and live in different neighborhoods. The variation in neighborhoods may have interfered with the environmental perception illustrated in the drawings, since each location has different problems.

FINAL CONSIDERATIONS

It can be seen that the students have two views of the environment (naturalistic and globalizing) and that the school lacks the ability to work on environmental education in an interdisciplinary way, since only three subjects work from this perspective: Biology, geography and history.

CHAPTER 4

PLANNING A PEDAGOGICAL INTERVENTION: CONTRIBUTIONS TO INITIAL TEACHER TRAINING UNDER PIBID

Alef Nunes da Silva[1] Auta Paulina da Silva Oliveira[2]

Màrcia Adelino da Silva Dias[3]

Abstract: In the era of considerable transformations and advances in science and information techniques, teacher training has been taking its place among the dialogues about education in the current social context, as something of extreme necessity for public schools that face challenges in the teaching-learning process, both structural and in terms of school management and teacher training. It was with the aim of socializing a set of information about the importance of initial teacher training in the light of the Institutional Teaching Initiation Scholarship Program (PIBID) that this paper was prepared, reporting on a didactic experience regarding the planning of a pedagogical intervention, whose action plan was entitled WEBiology. The educational intervention is associated with the PIBID biology sub-project at the State University of Paraiba (UEPB) and is being implemented at the Dr. Hortêncio de Sousa Ribeiro State School. The aim was to justify the relevance of programs aimed at teacher training, especially initial training, based on the experience of planning a pedagogical intervention, an essential stage in the organization of pedagogical work in the classroom. From the elaboration of the action plan for a pedagogical intervention, it was clear how important it is to develop the planning of didactic interventions in order to introduce undergraduate students to the current school context, so that the experiences gained during their participation in PIBID are reflected in improvements in the quality of education, and also in the improvement of current methodologies through the continued training of primary school teachers.

Keywords: Teacher training; Pedagogical intervention; Planning.

INTRODUCTION

In the era of considerable transformations and advances in science and information techniques, the issue of teacher training has been taking up a position among the dialogues about education in the current social context, as something of extreme necessity for public schools that face challenges in the teaching-learning process, both structural and in terms of school management and teacher training.

In order to develop a way of minimizing the problem, teachers must be bold and make use of the diverse knowledge they have acquired during their training. In addition, it is essential to make a collective effort to overcome the obstacles that stand in the way of building the quality education that is currently being debated and desired (LIMA; BARRETO; LIMA 2007).

For several decades now, teacher training in Brazil has been a major challenge for the democratization of education in the country and the rise in its quality, resulting in progress in the conditions of access and permanence of school-age children and adolescents in educational environments.

Through this knowledge, it is necessary to recognize the role of universities as training centers for future professionals, where the quality of teaching is linked to up-to-date and innovative working methods, as well as articulated with educational programs aimed at valuing teachers, such as the Institutional Teaching Initiation Scholarship Program (PIBID).

According to the CAPES website, the main objective of PIBID is to encourage the training of teachers at a higher level for basic education, helping to enhance the teaching profession through the inclusion of undergraduates in the daily life of[1] Graduated in Biological Sciences at the State University of Paraiba;[2] Master's student in the post-graduate program in agricultural sciences at the State University of Paraiba;[3] PhD Professor in the Biology Department at the State University of Paraiba.

schools in the public education system, providing them with methodological experiences that will improve the quality of their initial training and also promote links between higher education and basic education. Therefore, it contributes to the creation of innovative teaching practices that can help overcome the difficulties identified in the teaching-learning process.

However, PIBID is a program conceived by the Ministry of Education, which, in accordance with the legal requirements of the Coordination for the Improvement of Higher Education Personnel (CAPES), works to stimulate and promote the initial and continuing training of teaching professionals under Law No. 11,502 of July 11, 2007.

According to Delizoicov et al. (2007, p. 31):

"The challenges of the contemporary world, particularly those relating to the transformations that school education needs to undergo, have a direct impact on initial and continuing teacher training courses."

The training of education professionals doesn't end at graduation, quite the contrary, teaching is done by continuing and updating the knowledge acquired, and in this case, the extent of this concept must be analyzed, that is, the positioning and autonomy of teachers in reading the reality of their own training is indispensable; the situation of the public school and the channels that should emerge from reflections on this new look at the teacher training process, beyond the institutions.

Considering the function of planning the actions to be developed as a way of linking the reality of school and society to teacher training, the planning of educational tasks is a method of rationalizing, arranging and coordinating the teacher's actions: the elucidation of the principles, guidelines and methodologies of the work, as well as contributing to the organization and coordination of the teacher's actions, in order to establish coherence in the development of the plan, thus facilitating the preparation of lessons and measurement of results (LIBÂNEO, 1994).

It was with the aim of socializing a set of information about the importance of initial teacher training in the light of PIBID that this work was prepared, reporting on a didactic experience regarding the planning of a pedagogical intervention, whose action plan was entitled WEBiology, which seeks to recognize the cultural and revolutionary influences of society, such as the technological development of the media and social networks, which contribute to scientific thinking being improved and socialized, recognizing the relevance of the Internet in today's school reality.

The pedagogical intervention is associated with the PIBID biology subproject at the

State University of Paraiba (UEPB) and is being implemented at the Dr. Hortêncio de Sousa Ribeiro State School. The aim was to justify the relevance of programs aimed at teacher training, especially initial training, based on the experience of planning a pedagogical intervention, an essential stage in the organization of pedagogical work in the classroom.

METHODOLOGY

The pedagogical intervention WEBiologia was developed within the framework of PIBID/CAPES/UEPB to be developed in the 1ª grade, class B, of the High School of the State School of Innovative and Professionalizing High School Dr. Hortêncio de Souza Ribeiro located in the city of Campina Grande - PB.

This is a descriptive study in which an experience in the process of planning a pedagogical intervention will be reported, emphasizing the opportunity to participate in PIBID as a favorable aspect of the process of initial teacher training, taking into account, in this study, the development of learning in everyday school life based on autonomous participation in the process of preparing a pedagogical intervention proposal.

Initially, we experienced the daily life of the aforementioned school in order to see and identify the shortcomings and dilemmas faced by high school students in terms of their learning process, focusing on the teaching of biology.

Proposal for pedagogical intervention in biology teaching

ªThe stage of drawing up the pedagogical intervention proposal began after a visit to the primary school and after observing daily school life in order to detect possible difficulties in the teaching and learning process presented by secondary school students, especially in the teaching of biology.

Another important factor that led to this elaboration was the perception of why and how to draw up an action plan for a pedagogical intervention, with a view to seeing the role of the teacher as a reflective and autonomous professional in the elaboration of their educational project.

After these steps, we began to write an action plan entitled WEBiology, which was designed to introduce the use of the Internet in education into the students' daily lives, but precisely in the teaching of biology, which characterizes the pedagogical intervention in focus.

At this point, we followed a script drawn up by the PIBID supervisor at the school to plan the pedagogical actions we intended to develop, containing an objective description of the target audience, i.e. the class that would be intervened in, the contents that would be worked on and a convincing justification, so that the intervention proposal could be delimited.

The steps mentioned above allowed for a better way of measuring the results, after which the methodological path had to be described in detail in order to meet the initial proposal.

However, we intend to develop competencies and skills for teaching practice, by planning this pedagogical intervention and being able to evaluate the possibilities and limits found in the pedagogical action after it has been carried out.

RESULTS AND DISCUSSION

The PIBID in different subprojects plays a decisive role in initial and continuing teacher training, which for Saviani (2009) this issue of teacher preparation "arises explicitly after territorial independence from the Portuguese crown, when the organization of popular education is considered" and somehow "establishes itself as a possibility of articulation between theory and practice throughout the initial training process" (FELICIO, 2014, p. 419).

Programs like this are extremely important for the development of quality education in public schools, since they provide primary schools with "contact with the latest research generated by the University in the field of education and in the specific areas of each of the subprojects contemplated" (FERNANDES, MENDONÇA, 2013).

In accordance with the Education Guidelines and Bases Law (LDB) No. 9394 of September 20, 1996, which, among other aspects, clearly and specifically sets out the purpose of training education professionals. In this sense, it is worth highlighting the following points:

Art. 61: The training of education professionals, in order to meet the objectives of the different levels and modalities of education and the characteristics of each stage of the student's development, will be based on:

I - the association between theory and practice, including through in-service training;

II - drawing on previous training and experience in educational institutions and other activities.

Art. 62: The training of teachers to work in basic education will be done at a higher level, in a full degree course, in universities and higher education institutes, admitted, as a minimum training for the exercise of teaching in early childhood education and in the first four grades of elementary school, that offered at a secondary level, in the Normal modality. (BRASIL, 1996).

Still on the institutionalization of PIBID, it is worth noting that it has its roots in universities such as UEPB, which has been part of research and academic extension projects since 2009, to the detriment of Law 11.502, of July 11, 2007, which encourages and fosters the initial and continuing training of teaching professionals, in addition to meeting the guidelines of the All for Education Commitment Goals Plan through the principles set out in the National Policy for the Training of Teaching Professionals in Basic Education (Decree 6.755, of January 29, 2009 and Law No. 11.947, of June 16, 2009, in its art. 31).

In response to these PIBID determinations. Considering the challenges and dilemmas faced in the most varied determinations that condition the dissemination or transformation of society (LIMA; BARRETO; LIMA 2007), thus creating a teacher profile aware of its reality, which still according to the authors, this "requires the construction of a professional identity that is not immutable, but capable of in-depth readings of the educational phenomenon."

From this perspective, we can justify the use of the methodological tool chosen for this project, the internet, recognizing its relevance in today's school reality. This was achieved by visiting the school and observing its daily life, which allowed the target audience to be better analyzed, in order to contribute to the process of preparing the intervention.

The challenge encountered in observing everyday school life was the weak ability to interconnect the knowledge acquired at school with everyday life, which makes it difficult to socialize knowledge, which in turn prevents the information obtained in the classroom from being improved and contextualized.

According to Nonaka and Takeuchi (2008), socialization is the method in which intellectual and personal experiences are shared from person to person. It occurs through day-to-day activities in various situations and places, elaborating knowledge through direct experiences.

In an attempt to solve this problem, the pedagogical intervention was designed, making use of the media and social networks as a way of socializing learning.

Preparation of the intervention proposal

By drawing up the action plan, it was possible to reflect on the role of the teacher as a reflective professional who must be attentive to patterns of events, be able to describe what they observe, be inclined to propose daring and sometimes radically simplified models of experiments and ingenuity in proposing ways of testing them that are compatible with the limits of an action environment (SCHON, 2000).

However, in order for the teacher to have this reflective character, acting in the face of adversity, he or she must have the autonomy to plan and put his or her project into practice, or rather have control of his or her own work, which for Contreras (1997) is defined when applying teacher training:

[...] in short, a dynamic process of definition and personal constitution of who we are as professionals and of the awareness and reality that this definition and constitution can only be carried out in the context of confirming one's own professional reality, which is the encounter with other people, both in our mission to influence their personal training processes and in the need to define or contrast with other people and other sectors what this training should be like (CONTRERAS, 1997, p.63).

This is how teachers acquire "the *status* and ability to carry out their role with autonomy, responsibility and daring" (DASSOLER; LIMA, 2012, p.7). This is explained by the teacher's ability to objectify their work, which is perceived when they characterize what can be achieved with it, through the skills and competences that can be acquired by the student.

Although it is not considered a focus of discussion in this study, it is important to emphasize that the interaction between undergraduates and teachers, integrated into PIBID, provides favorable conditions for the continued training of public school teachers, based on relationships with teachers in training, which Dassoler and Lima (2012, p.7) call:

"constantly educating oneself through learning in which the knowledge constructed results in new relationships with other knowledge which, in turn, generates new constructions." This shows that the reciprocity of experiences between teachers and undergraduate students certainly contributes significantly to the exchange of knowledge, enriching current methodologies with up-to-date didactic proposals.

The reflection and autonomy described above as fundamental characteristics for teacher training, was perceptible when drawing up an action plan, because some skills and observations of the school were essential for drawing it up.

The perception of the school's reality and also of the students' behavior in the face of the resources made available by the school, undoubtedly helped us to intervene with constructive proposals for learning, using innovative methods, which added to the experience of teaching knowledge, contributing to our training as education professionals.

FINAL CONSIDERATIONS

From the preparation of the WEBiology action plan in compliance with the requirements of the Institutional Teaching Initiation Scholarship Program, I realized how important it is to develop planning for didactic interventions to introduce undergraduate students into the current school context, so that the experiences gained during participation in the PIBID are reflected in improvements in the quality of education, and also in the improvement of current methodologies through the continued training of primary school teachers.

As an initial teacher training program, PIBID makes a significant contribution to expanding the quality of initial teacher training in undergraduate courses and also to the continuing training of primary school teachers, based on relationships with teachers in training.

The program is capable of creating bridges between the educational institutions involved in the construction of education professionals and also in the creation and expansion of teaching methodologies that essentially seek interdisciplinarity in academic and social requirements.

It is also active in creating a teacher profile that is aware of its reality, and autonomous in the development and application of its educational projects, which, upon realizing the challenges found in the school environment, seeks to solve this problem by discussing the problem, and putting into practice ideas to end it and at the same time improve teaching-learning relationships.

From this perspective, planning pedagogical work is extremely important, because a teacher who plans well will be better able to monitor the development of students' learning and this will certainly contribute to the quality of teaching, increasing students' interest in classes, thus assuming an essential stage in the organization of school work.

BIBLIOGRAPHICAL REFERENCES

BRAZIL. Decree No. 6.755, of January 29, 2009. Establishes the National Training Policy for Basic Education Teaching Professionals, regulates the work of the Coordination for the Improvement of Higher Education Personnel - CAPES in fostering initial and continuing training programs, and makes other provisions. **Official Gazette of the Federative Republic of Brazil.** Brasilia, January 29, 2009. Available at: <http://www.planalto.gov.br/ccivil_03/_ato20072010/2009/decreto/d6755.htm>. Accessed on: 05 Jun 2015.

. Law No. 11.502, of July 11, 2007. Modifies the competencies and organizational structure of the Coordination for the Improvement of Higher Education Personnel - CAPES foundation, referred to in Law No.° 8.405, of January 9, 1992; and amends Laws No.os 8.405, of January 9, 1992, and No. 11.273, of February 6, 2006, which authorizes the granting of study and research scholarships to participants in initial and continuing teacher training programs for basic education. **Official Gazette of the Federative Republic of Brazil.** Brasilia, July 11, 2007. Available at:<http://www.planalto.gov.br/ccivil_03/_Ato20072010/2007/Lei/L11502.htm>. Accessed on: June 4, 2015.

. Law No. 11.947, of June 16, 2009. Provides for the provision of school meals and the Money Directly at School Program for students in basic education; amends Laws No.os 10.880, of June 9, 2004, No. 11.273, of February 6, 2006, No. 11.507, of July 20, 2007; repeals provisions of Provisional Measure No. 2.178-36, of August 24, 2001, and Law No.° 8.913, of July 12, 1994; and makes other provisions. Brasilia, **Diàrio Oficial [da]**

Repùblica Federativa do Brasil. 16 June 2009. Available at: <http://www.planalto.gov.br/ccivil_03/_ato20072010/2009/lei/l11947.htm>. Accessed on: 06 Jun 2015.

BEZERRA, T. M. O.; GONÇALVES, A. P. C. **Concepções de meio ambiente e educaçao ambiental pelos professores da Escola Agrotécnica Federal de Vitória de Santo Antao-PE.** Biotemas. Jaboatao do Guararapes, v.20, n.1, p. 115-125, 2007.

BEZERRA, T. M. O.; FELICIANO, A. L. P.; ALVES, A.G.C. Percepçao ambiental de alunos e professores do entorno da Estaçao Ecológica de Caetés- Regiao Metropolitana do Recife-PE. **Biotemas**. Recife, v.21, n.1, p. 147-160, 2008.

BRAZIL. Law n. 9.394, of December 20, 1996. Establishes the guidelines and bases of national education. **Official Gazette of the Federative Republic of Brazil**. Brasilia,20 de Dec. 1996. Available at: <http://www.planalto.gov.br/CCIVIL_03/leis/L9394.htm>. Accessed on: 04 Jun 2015.

CARVALHO, I. C. M. What is environmental education? Elements for a debate on environmental education and rural extension. **Agroecologia e Desenvolvimento Rural Sustentàvel,** Porto Alegre, v.2, n.2, p. 43-51, 2001.

CONTRERAS, J. La Autonomia Del Profesorado. Madrid: **Morata**, 1997.

DASSOLER, O. B; LIMA, D. M. S. Teacher training and professionalization: characteristics, daring and knowledge. In: IX ANPED SUL Seminar, 2012, Caxias do Sul. **Electronic proceedings of the IX ANPED SUL**, Caxias do Sul. UCS, 2012. Available at:<http://www.ucs.br/etc/conferencias/index.php/anpedsul/9anpedsul/paper/viewFile/3171/522>. Accessed on: June 10, 2015

DELIZOICOV, D.; ANGOTTI, J. A.; PERNAMBUCO, M. M. Ensino de ciências: fundamentos e métodos. 2 ed. Sao Paulo: **Cortez**, 2007.

FERNANDES, M. J. S; MENDONÇA, S. G. L. PIBID: A contribution to teacher training policy. **EntreVer**, Florianópolis, v. 3, n. 4, jan. 2013. Available at: <http://incubadora.periodicos.ufsc.br/index.php/EntreVer/article/view/2581/3081>. Accessed on: June 10, 2015.

LIBÂNEO, J. C. Didâtica. Sao Paulo: **Cortez**, 1994

LIMA, P. G; BARRETO, E. M. G; LIMA, R. R. Teacher training: a necessary reflection. **Educere et Educare**, Paranà, v. 2, n.4., 2007. Available at: <http://erevista.unioeste.br/index.php/educereeteducare/article/viewArticle/1657>. Accessed on: 03 Jun 2015.

NONAKA, I; TAKEUCHI, H. Gestao do Conhecimento. Porto Alegre: **Bookman**, 2008.

ROSA, L. G.; LEITE, V. D. ; SILVA, M. M. P. Conception of environment and environmental education of educators from a school of initial training in pedagogy, secondary level. **Revista. eletrônica de Mestrado em Educaçâo Ambiental**. Rio Grande, v.18, p. 244-259, 2007.

ROSA, L. G.; SILVA, M. M. P.; LEITE, V. D. Environmental education in a high school: strategies and challenges in the sensitization process. **Revista Eletrônica de**

Mestrado em Educaçâo Ambiental. Rio Grande, v.22, p. 454-475, 2009.

REBOLLAR, P. M; Environmental education and the terms environment and environmental impact in the view of higher education students in the greater Florianópolis region - SC. **Biotemas.** Florianópolis, v.22, p.173-180, 2009.

ERROBIDART, N.C.G; GOBARA, S.T.; JARDIM, M.I.A; ERROBIDART, H.A; MARQUES, S.M.; Mental models and representations used by high school students to explain waves. **Revista Electrónica Enseftanza de las Ciências**,v. 12, n.3, p.440-447, 2013.

FREIRE, P. Pedagogy of the oppressed. 17°ed. Rio de Janeiro. **Paz e Terra**.1987.

GIORDAN, M. Experimentation and science teaching. **Quimica nova na escola.** Sao Paulo, v.10, p. 43-49, 1999.

MIZUKAMI, M.G.N. **Ensino as abordagens do processo**. Sao Paulo, EPU. 1986.

MORAES, R.(Org); ROSITO, B.A.; HARRES, J.B.S.; GALIAZZI, M.C.; RAMOS, M.G.; COSTA, R.C.; BORGES, R.R. **Construtivismo e ensino de ciências reflexôes epistemológicas e metodológicas**. Porto Alegre, EDIPUCRS. 2000.

PAGEL, U.R.; CAMPOS, L.M.; BATITUCCI, M.C.P. Teaching methodologies and practices: A reflection on the contribution of practical classes in the biology teaching-learning process. Mato Grosso. **Experiência em ensino de Ciências**,v.10, n.2, p. 14-25, 2015.

NEVES, M. S.; CABALLERO, C.; MOREIRA, M.A. Rethinking the role of experimental work in physics learning in the classroom - An exploratory study.

Portugal. **Investigaçôes em Ensino de Ciências**. V. 11, n.3, p. 383-401, 2006.

BEZERRA, T. M. O.; GONÇALVES, A. P. C. Concepções de meio ambiente e educaçao ambiental pelos professores da Escola Agrotécnica Federal de Vitória de Santo Antao-PE. **Biotemas**. Jaboatao do Guararapes, v.20, n.1, p. 115-125, 2007.

BEZERRA, T. M. O.; FELICIANO, A. L. P.; ALVES, A.G.C. Percepçao ambiental de alunos e professores do entorno da Estaçao Ecològica de Caetés- Regiao Metropolitana do Recife-PE. **Biotemas**, Recife, v.21, n.1, p. 147-160, 2008.

CARVALHO, I. C. M. What is environmental education? Elements for a debate on environmental education and rural extension. **Agroecology and Sustainable Rural Development**. Porto Alegre, v.2, n.2, p. 43-51 2001.

ROSA, L. G.; LEITE, V. D. ; SILVA, M. M. P. Concepção de ambiente e educaçao ambiental de educadores e educadoras de uma escola de formação inicial em pedagogia, nivel mèdio. **Rev. eletrônica de Mestrado em Educaçâo Ambiental**, Rio Grande, v.18, p. 244-259, 2007.

ROSA, L. G.; SILVA, M. M. P.; LEITE, V. D. Environmental education in a high school: strategies and challenges in the sensitization process. **Rev. Eletrônica de Mestrado em Educaçâo Ambiental**. Rio Grande, v.22, p. 454-475, jan/july, 2009.

REBOLLAR, P. M; Environmental education and the terms environment and environmental impact in the view of higher education students in the greater Florianópolis region - SC. **Biotemas**, Florianópolis, v.22, p.173-180, 2009.

SCHON, D. Educating the Reflective Professional: a new design for teaching and learning. Porto Alegre: **Artmed**, 2000.

SOARES, L. G. C; SALGUEIRO, A. A; GAZINEU. M. H. P. Environmental education applied to solid waste in the city of Olinda, Pernambuco - a case study. **Revista de ciência e tecnologia**, Recife, v.1, p.1-9, 2007.

MIX
Papier aus verantwortungsvollen Quellen
Paper from responsible sources
FSC® C105338

Printed by Books on Demand GmbH, Norderstedt / Germany